Fire can engulf a building in minutes

ELEMENTS

Aaron Frisch

A+
Smart Apple Media

COPYRIGHT

Published by Smart Apple Media
1980 Lookout Drive, North Mankato, MN 56003
Designed by Rita Marshall

Printed in the United States of America
Photographs by Frank Balthis, Pat Berret, Horticultural Photography, Tom Myers, Erwin "Bud" Nielsen
Library of Congress Cataloging-in-Publication Data
Frisch, Aaron. Fire / by Aaron Frisch. p. cm. – (Elements series)
Includes index.
ISBN 1-58340-075-3
1. Fire–Juvenile literature. [1. Fire.] I. Title. II. Elements series (North Mankato, Minn.)
TP265 .F78 2001 541.3'61–dc21 00-054170
First Edition 9 8 7 6 5 4 3 2 1

CONTENTS

Fire

What Is Fire?

Humans probably first discovered fire when lightning struck a tree. The fire kept them warm, gave them light, and let them cook animals for food. But fire was very hard to start, so many ancient peoples kept public fires burning at all times in every village. Nothing was more important than their fires. Fire is the heat and light that come from burning substances. A fire happens when oxygen—a common gas in the air—combines very quickly with another substance. This rapid

Fires can start when lightning strikes

combination is called a chemical reaction. The faster the oxygen joins with the substance, the hotter and brighter the fire gets. Three things are needed to make fire. The first one is a fuel—a substance that will burn. Fuel is like food for the fire. It can be solid (such as wood), liquid (such as gasoline), or gas (such as natural gas). The second thing needed to make fire is heat. If the fuel does not get hot enough, it will not burn. The third thing needed to make fire is oxygen. Oxygen is usually available in the air surrounding

Not all substances produce flames when they burn. Charcoal, for example, just glows brightly and gives off heat.

the fuel. Without any one of these three things, a fire cannot burn. To put out a campfire, for example, campers need simply to take away one of the three things. They may pour

Trees, especially dead ones, are a good source of fuel

The oxygen in air makes fire possible

water on the fire to take away the heat. They may kick sand on the fire to take away its oxygen. Or they may stop adding wood to take away its fuel.

Building Heat

Some fuels need more heat to burn than other fuels do. Solid fuels need to be much hotter than liquids or gases before they will burn. Most kinds of wood and plastic must be between 500° F and 900° F (260° to 480° C) to start on fire. But gasoline can start on fire even when it is as cold as –36° F (–38° C). Fires sometimes start by themselves. This is

called **spontaneous combustion**. This can happen when people throw away rags that are soaked in oil. Over time, oxygen unites with the oil, and heat builds up inside the rag. If

Forests provide ample fuel for wildfires

the rag gets hot enough, it can suddenly burst into flames.

Nothing can resist fire. Even concrete and steel can be cracked or melted by the heat of a very intense fire. Still, materials can be treated so that they are harder to burn. **Flame retardants** are often added to clothes, carpets, and building materials so that they are harder to start on fire and do not burn well.

The flames of a fire are usually orange or blue. The color depends on the substance that is burning and how hot it is.

Planes may drop fire-retardant chemicals to smother wildfires

Fuel for Fire

Matches were invented in 1827. Before that time, starting a fire was hard work. People would gather small piles of **tinder**. Then they would strike a hard rock against a piece of steel to make sparks that fell into the tinder and started it on fire. People also started fires by twirling a stick in a small notch in some other wood. This made sawdust that slowly got hotter until it produced flames. Fire does not burn all of the fuel it uses.

The tips of matches are covered by chemicals that need just a little heat to start on fire. This heat is created by rubbing them on a rough surface.

Most kinds of fuel contain some **minerals** that will not combine with oxygen, and so won't burn. Ashes, like those left in

Ashes pile up in a wood-burning fireplace

a fireplace after burning wood, are produced by these unburnable minerals. Fire may also produce smoke. Smoke is a mixture of **soot** and gases produced by fire.

Fire Uses and Dangers

Fire helps people in many ways. Early humans relied on fire to stay warm, cook food, and make light. People still use fire for these things today, but it has other uses, too. By burning gasoline, fire today is used to power cars, planes, machines, and many other things. Fire is also used to destroy garbage

Burning chemicals produce spectacular fireworks

and kill **germs**. Many people find it relaxing to watch and listen to campfires. Fire helps people in many ways, but it can also be harmful. Fires that get out of control are very dangerous. They can kill many people and destroy anything in their path, including forests, homes, and buildings. Still, fire helps us in many more ways than it hurts us. As long as people respect its power, fire is a wonderful tool of nature.

An explosion is the very rapid burning of a fuel such as gunpowder or dynamite.

Fires claim many lives each year

Candle Vapor and Heat

Things that are already hot are easier to start on fire than things that are cold. You can see this for yourself with a simple experiment. (Ask an adult to help you with this activity.)

What You Need

A candle

Matches

What You Do

1. Strike a match and light the candle. Let it burn until the wax around the wick has melted.
2. Blow out the candle. You'll notice that smoke rises from the wick after the flame goes out.
3. Strike another match and hold its flame in the rising smoke. The flame from the match will quickly travel down the smoke and relight the candle.

The candle relights because the smoke that rises up is made of wax that has vaporized. Because this vaporized wax is still hot, the flame from the match starts the vapor on fire. The fire then travels down the vapor to the hot wick and lights it.

Many people burn candles as decorations

INFORMATION

Index

Words to Know

flame retardants (FLAYM ree-tar-dents)—chemicals added to substances to make them harder to burn

germs (JERMS)—tiny organisms that can cause diseases

minerals (MIN-uh-ruls)—basic particles found in rocks and other non-living substances

soot (SUHT)—tiny, unburned particles left over after a fire

spontaneous combustion (spon-TAY-nee-us cum-BUS-chun)—the lighting and burning of a fuel without the help of outside heat

tinder (TIN-der)—materials that burn easily, such as grass, tree bark, and cotton

Read More

Landau, Elaine. *Fires.* New York: Franklin Watts, 1999.

Patent, Dorothy Hinshaw. *Fire: Friend or Foe.* New York: Clarion Books, 1998.

Ritchie, Nigel. *Fire.* Brookfield, Conn.: Copper Beech Books, 1998.

Internet Sites

Fire Safety
http://www.firesafety.org/index.cfm

FireNet
http://www.fire.org.uk/

United States Fire Administration Kids Page
http://www.usfa.fema.gov/kids/